BRING THE

Wild

INTO YOUR GARDEN

BRING THE WILD INTO YOUR GARDEN

An Hachette UK Company
www.hachette.co.uk

Summersdale Publishers Ltd
Part of Octopus Publishing Group Limited
Carmelite House
50 Victoria Embankment
LONDON
EC4Y 0DZ
UK

www.summersdale.com

Printed and bound in China

ISBN: 978-1-78783-667-9

Substantial discounts on bulk quantities of Summersdale books are available to corporations, professional associations and other organizations. For details contact general enquiries: telephone: +44 (0) 1243 771107 or email: enquiries@summersdale.com.

BRING THE

Wild

INTO YOUR GARDEN

ANNIE BURDICK

summersdale

Contents

Introduction

When we garden, we give a little bit of ourselves back to nature. We crouch in the soil, we dig our hands deep into the earth and we cultivate plants and produce. We become more in tune with the earth's rhythms and cycles, and we allow nature to run its course, unable to take control over it. Connecting to nature in this way can be deeply grounding, and it can offer us some useful perspective when we feel caught up in the usual commotion of our daily lives.

But there's more to gardening than planting seeds and watering your flower beds. There's an endless world of wildlife out there for you to discover and appreciate. Though gathering seeds and taking great care to plant them may feel like a conclusively satisfying accomplishment, why not go a little further by making your outdoor space a haven for the wildlife that surrounds you.

Introducing wildlife into your garden is endlessly rewarding. With just a little time and energy (and the right supplies), you can transform your garden or outside space into a place which sustains and supports some living inhabitants; a tiny, functioning ecosystem, with its own small-scale food chain.

On top of all this, increasing the biodiversity of your garden is a great way to support local wildlife, providing a safe space for insects and animal life to survive and thrive.

Ultimately, this book is for any gardener. Whether you have a tiny balcony or a sizeable plot, a quaint courtyard or a huge back garden, you'll see how simple it can be to become a host and friend to local wildlife. All you have to do is assess your space and your goals, then get creative.

Make wildlife-friendly gardening work for you. When you can step out your door and see birds stopping by for a bath, butterflies drifting through the air, bees coming in for a meal, or frogs warming up in the sun, you'll realize that the work was worth it.

If you do good to nature, it will reward you with its beauty every time.

CHAPTER 1

Birds

Making your garden more bird-friendly is an easy place to start your wildlife garden adventure. Not only are birds among the most conspicuous of garden wildlife, but they also tend to be fairly easy to attract.

Bird food is readily available and bird feeders are easy to make at home; bird baths are also easy to find and make for a fantastic, cheap weekend upcycling project. Or, if you're looking for a slightly longer-term investment, you could introduce bird houses or fruit trees into your garden.

If you have a larger garden with some established trees, it's likely that birds are already stopping by. This gives you a head start, and adding just a few home-made bird feeders or a bath will be enough to prove to these little guests that they should come by more often.

Still, if you're working with a smaller space, perhaps a balcony, courtyard, or deck, don't worry. With some care and a little patience, small spaces can still play host to visiting birds who will appreciate the effort you've made to welcome them in.

Seeing birds in your garden is a great sign that you're beginning to create a small-scale ecosystem in your own outdoor space. Birds are an important part of the food chain and their presence also often signals the arrival of other creatures, including all manner of insects and even small mammals.

In every walk with nature

one receives far more

than he seeks.

JOHN MUIR

Vary Your Feeders

With the massive variety of bird feeder styles and sizes available, choosing only one for your garden might feel like a daunting task. Thankfully though, it's actually best to use more than one type of feeder, to cater to a wider variety of local birds. Smaller birds often have an easy time perching on even delicate feeders, and their smaller beaks may only be able to select certain smaller seeds. Meanwhile, larger birds may be too heavy for some feeders and prefer sturdier varieties that will support them and offer up plenty of food. So, instead of picking out one type of bird feeder – only to discover that your neighbourhood birds aren't interested – select two or more different types and try putting different foods in each. This will make your garden more attractive to a diverse group of winged visitors.

Cater to Your Birds

Common bird species can vary greatly even from town to town or neighbourhood to neighbourhood. But, in general, the most common bird species for your climate and area are likely to be the most frequent visitors to your garden, depending of course on the feeders and food you choose.

There are several main types of bird feeders, each with its own benefits and considerations. Here are some things to keep in mind about each:

- Platform feeders: these are essentially open trays, mounted on a pedestal or pole. These large feeders attract the widest variety of birds (and possibly the greatest quantity as well). They also provide more of an open invitation for squirrels and – if low to the ground – other small animals to stop by for a bite.

- Hopper feeders: these are enclosed containers that have a small opening for the birds to feed from; they do a better job of protecting the food and seeds from the elements, but can be harder to clean out. They generally attract "feeder" birds, such as finches, sparrows and jays.

- Window feeders: these are usually similar in structure to platform feeders and attach directly to a window (often with suction cups), and provide the best close-up birdwatching opportunities. They're also easier to clean and check in on.

- Tube feeders: these are long, hollow tubes which tend to keep food dry, but also cater specifically to small birds and those who feed while perched upside down.

Placing Your Feeder

Beyond choosing the right size and style for your requirements, perfecting the placement of your bird feeder is critical to attracting local birds into your outside space. The priority should always be to place any feeder in a spot that will be safe for the birds; if they ever feel afraid at your feeder, then they aren't likely to return.

With this in mind, avoid overly noisy or wide-open areas, as this will leave birds feeling exposed and vulnerable to threat. Avoid placing your feeder too low to the ground, as birds are more likely to be threatened by a cat or other predator in this position. Finally, although it may initially seem counterintuitive, consider using a window feeder to reduce the risk of birds flying into your windows; the feeder will give them a better visual representation of where the glass is located.

Keeping Your Feeder Clean

Most feeder types are prone to letting in at least some moisture, which can lead to the food becoming mouldy or rotten over time. Although platform feeders are particularly prone to this (being open and exposed to the elements), even covered feeders will need to be cleaned and refilled regularly.

As rotting seeds and husks can make birds sick, it's important to be vigilant about cleaning your feeders. The simplest method for cleaning is to separate all pieces of the feeder and wash them in a dishwasher – or handwash using soap and boiling water (or a highly diluted bleach solution).

The last step for keeping your bird feeding area clean – and the easiest to forget – is to clean the ground around your feeders. Seed husks and bird food will often drop to the ground, and there they can also become mouldy, potentially causing illness to any creatures that consume them.

Choosing the Right Bird Food

An error that many people make when they're new to feeding birds is to buy food for the birds they *want* to attract – not the ones that are already there. Birds, as social creatures, learn from each other and gather in groups. So, the first step in choosing a specific food is to try to please the birds you know are already around; once these birds start to stop by regularly, other new birds will take note and come to your feeders as well. Then you can safely add new feeders and food that you know will be a better match for your new guests. Try to also think about the places you're purchasing your bird food from. Local garden shops are best, since they will know about local species and offer reputable produce.

Being Hospitable

Another way you can provide a great garden environment for birds is to include a nesting box in your outdoor space. While some birds prefer to establish their nest in grassy areas or dense brush, others look for a more enclosed space in which to lay eggs and seek shelter. Providing a nest box for your neighbourhood birds can help support local species populations and bring more of the wild into your garden.

When installing a box, ensure that it:

- Is made of untreated wood

- Has thick walls to keep it insulated

- Has an entrance hole of practical size (which is a good distance from the bottom of the box)

- Includes drainage and ventilation holes

- Has a slanted roof to keep rain out

- Doesn't have exterior perches (these can provide easier access for predator birds)

DIY Bird Feeders

Ready to take matters into your own hands? A simple option is to make bird-feed fat balls. Home-made fat balls mimic the suet bird-feed blocks you can find in most garden shops, but cost a fraction of the price.

All you need is lard or suet (either vegetable or beef) and bird-safe seed mix. Combine two-parts seed with one-part suet and heat slowly in a pan, stirring until they've melted together. Scoop the mixture and roll into balls, then freeze to harden them. These fat balls can be placed in any empty hanging feeder or on a platform.

If you want to go a step further, you could scoop your suet mixture into empty plastic cups or containers before you freeze the mixture. Once they're ready, simply attach twine or string and hang them at an appropriate height.

Another option for a DIY feeder is to create your own bird-seed holder from an empty plastic bottle or jug (making sure to fully wash the container first). First, cut a hole around the middle of the bottle, large enough that a small bird could get in. Then, make two small holes below the hole you have just made on each side of the bottle; slide a wooden stick or dowel through from one to the other to create a perch. Fill the bottle with bird seed to below the point of the larger hole and hang in a smart spot for birds to access.

What wild creature
is more accessible
to our eyes and
ears, as close to
us and everyone
in the world, as
universal as a bird?

DAVID ATTENBOROUGH

Choosing the Right Bird Bath

Beyond stocking a bird feeder or two, you can make your garden even more appealing to birds by adding in a bird bath. Like bird feeders, these come in a multitude of styles and sizes, so it helps to keep a few things in mind when you're choosing one. If you have a small outdoor space, try repurposing a small water dish or candle plate. If you have slightly more room, investing in a larger, elevated bath will help attract a multitude of birds to your garden.

Next, consider bird comfort: a narrow edge on your bird bath will be easier for small claws to hold, and a depth of 3–5 cm is ideal for this. Placing your bird bath some distance from the ground will also ensure your bird visitors are safe from predators.

Finally, think about how easy it is to clean your bird bath. Ideally, look for a bird bath with a bubbling or dripping feature, as moving water will help with cleaning maintenance and help prevent mosquitoes breeding.

Home-Made Bird Baths

At its core, a bird bath is simply a shallow dish filled with water, attached to a stand of some sort (set at an appropriate height). This relatively flexible brief means that making your own bird bath – either by upcycling unused household items or heading to a vintage shop – is quite straightforward. The possibilities are endless, but here are a few ideas you can consider for the "bath" element of your home-made bird spa:

- Old serving dishes
- Metal bin lids
- Glassware (bowls, dishes, platters and the like)
- Ceramic cake pans

As for the stand, you can be creative here too – just as long as you attach the two parts together safely. You could try:

- A painted step stool
- Stacked bricks
- A lamp base
- Metal plant stands
- Placing the bath on a wall, table or other stable garden feature

Birds and Berries

Having the right seed, feeders and baths is a great start to making your garden a bird-friendly spot. To take it to the next level, consider that many bird species prefer to eat fruit over seeds. To continue diversifying the creatures you attract to your garden or outdoor space, a berry- or fruit-bearing plant is a great step. Plus, with a large enough plant, you could see your outdoor space become a permanent nesting place for local birds, since many species like to nest in bushes and trees.

What's more, many bushes and trees which bear fruit safe for birds are also suitable for humans. Want to feed yourself and wildlife at once? Apricot, orange, plum, apple, pear, cherry, peach or raspberry plants are all great options.

CHAPTER 2

Butterflies

and

Moths

Butterflies come in an array of lovely colours, sizes, patterns and shapes, and for many they conjure images of sunshine, warmth and blooming flowers. They catch our eyes and charm us with their flight. But, whether you long to attract them for their beauty, or you'd like to do your part to provide habitats for these declining species, this chapter will show you how to make your garden into a butterfly haven.

Unfortunately, butterflies are more sensitive than most creatures to climate change, so as time goes on, these fragile insects are being lost in great numbers. Doing what you can to create safe butterfly spaces and butterfly food sources is a great way to give back to nature. And besides, butterflies do plenty of good as well as being lovely to look at! They're pollinators and are particularly integral in pollinating some varieties of flowers, including alyssum, milkweed, lilac and calendula. Some species are predators in their caterpillar stages, munching on insect species like aphids.

Ultimately, seeing butterflies in your garden means that your environment is thriving; once you put in the effort to create a space for them, they'll be back to enjoy your garden time and time again.

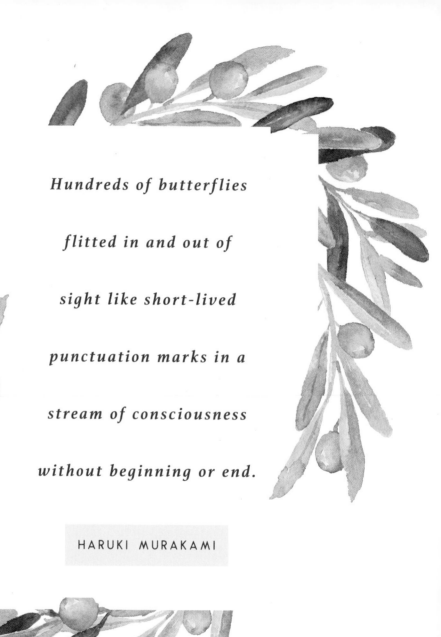

Hundreds of butterflies

flitted in and out of

sight like short-lived

punctuation marks in a

stream of consciousness

without beginning or end.

HARUKI MURAKAMI

Sun Seekers

Unless butterflies are constantly moving or sitting directly in sunlight, their bodies won't stay warm. This is why you're likely to see butterflies spreading their wings in sunny spots and flitting away from shade.

The first step to any successful butterfly hospitality in your own garden, then, is to have plants which attract butterflies in the sunniest possible spots. If you're not sure where your garden's best sunspots are, spend a day at home checking on your outdoor space every so often, making note of sunny patches. The patches that get the most light throughout the day are great candidates for your nectar-rich flowering plants. Just make sure those plants are also suited to lots of sunlight!

Get Planting

If you have a small outdoor space, creating a fully fledged butterfly garden with a variety of plants may seem unrealistic. However, the goals you have for your garden space don't have to be the same as for someone with a huge amount of land with plenty of flowering bushes. There are also easy ways to DIY and refurbish containers into planters to save you space and money.

Here are some simple tips for a small-scale butterfly garden:

- Choose your containers! Great and affordable options include: baskets, terracotta pots, wooden boxes, sieves, repurposed tyres and metal canisters. In fact, you can be endlessly creative. (Many nectar-producing plants need space to sprawl, so just keep size in mind when matching plant to container.)

- If your container is overly porous – like a basket that will let dirt escape – line it with a plastic bag first, cutting several slits in the bottom for water to drain through.

- If you're starting with seeds, plant directly into the containers, with specific flowers chosen for each planter. Seedlings should appear within one to two weeks, and flowers in as little as a month (depending on blooming season).

- If you'd rather get a head start, you can purchase plants and skip the seedling phase. This could be less satisfying, but is also quick and low-risk! You'll have flowers right away.

Be imaginative with your garden space. Hang planters, stand them on an old step-stool or other firm surface, or cluster pots together for a prettier look.

Planting Smarter

Ready to get your butterfly garden blooming? Aim for plants that flower for much of the year and provide good sources of nectar. If you do your research, you can plant a variety of flowers in early spring that will keep blooming through the autumn. Here are a few ideal starter plants to entice nearby butterflies:

- Common lilac: a beautiful shrub with a sweet, heady scent when it blooms in the late spring and early summer.

- Buddleja: a popular flower for adult butterflies, the plant will bloom for several weeks at a time.

- Verbena: a great source of nectar, these beautiful purple flowers bloom later in the season.

- Hebe: a hearty shrub that's beloved by both butterflies and bees.

- Hemp-agrimony: featuring thick stems and lovely light pink blooms, these plants thrive in damp areas and climates.

- Bowles's Mauve: frequently deadhead this plant and it's likely to bloom all spring, summer and autumn.

Keep Things Fruity

Late-season butterfly species are often much keener to snack on ripened fruit than flowery nectar. Of course, late in the season is exactly when fruit trees are loaded with fruit and dropping their excess, ripe fruit to the ground. If you're lucky enough to have fruit trees in your space and have picked your share of it by the end of the summer and into the autumn, leave fallen fruit on the ground. Ripe – even rotting – fruit, like apples, pears and peaches, are an ideal food source for many butterflies. Leaving fruit on the tree or on the ground will give them another reason to visit your garden for a tasty, sweet meal.

Hospitality Tips

On the lookout for other easy ways to make your garden the ideal place for a butterfly stopover? Here are some simple tips:

- Provide puddles: butterflies seek water for hydration and important minerals. Make things easier for them by placing flowering plants near spots with shallow puddles. If this isn't an option, make your own DIY "puddle" by leaving a very shallow dish with pebbles or sand and water in your garden. If it's close to nectar-producing plants it's likely to become a butterfly watering hole.

- Share your fruit: with a small garden space, you may not have a fruit tree, but you can still leave small pieces of extra-ripe fruit in a dish in your outside space. Replace it often, though, if you don't want ants to take over.

- Add molasses or fermented beer: these syrupy delights, added to a plate of fruit, will make for an enticing butterfly lunch.

Avoid Deterrents

Though there are plenty of ways to attract and care for visiting butterflies, it's also important to consider factors that could be keeping them at bay. Firstly, avoid pesticides completely. Butterflies and moths are very susceptible to pesticides – even some of the organic options. They can easily kill these delicate creatures, or otherwise disrupt their eating and mating patterns. Secondly, avoid "butterfly houses"; you might come across them online or at some garden shops. While there's a chance they may shelter a butterfly from time to time, the much greater possibility is that paper wasps will turn it into a convenient nest and it won't help butterflies at all. A small pile of wood and logs in a sheltered part of your garden will make a much better resting place for butterflies and all manner of other wildlife.

Cooking for Moths:
Sugaring Recipes

Not looking to limit your flying visitors to birds and butterflies? Moths also make for beautiful garden visitors, and mostly visit late in the evening and at night. However, as opposed to being drawn by nectar, moths are enticed by syrupy, sugary "traps" and fermenting household foods. Start with this basic recipe for a moth sugar bait:

Ingredients

Tin of black treacle (or molasses)
1 kg brown sugar
500 ml brown ale

Method

1.   Slowly heat the ale in a pan and simmer on a low heat for about five minutes.
2.   Stir in the brown sugar until it dissolves, then add the treacle and let it do the same. Simmer for another couple of minutes.
3.   Allow the mixture to cool.
4.   Around dusk, use a paintbrush to spread the sugar mixture onto trees, poles or fence posts (this process is called "sugaring"). Make palm-sized squares around eye level for the best view of the moths coming to eat. Doing this at dusk makes the mixture fresher when moths come out to feed, making it more appealing to them.

Here are some other items that make great additions to a sugar bait mix:

- Rum or sweet liqueurs

- Honey or maple syrup

- Dry yeast (particularly if you're letting your mixture sit and ferment)

- White sugar

- Mashed overripe bananas

- Rotting watermelon or other fermenting fruit like pears, apples and peaches

- Cola (simmered with the other ingredients so it dissolves)

Adopt the pace
of nature:
her secret
is patience.

RALPH WALDO EMERSON

Wine Ropes

Another great and cost-effective method for drawing moths to your garden is to hang wine ropes (long strips of fabric, soaked in sweetened wine). Here's a quick DIY you can do at any time:

1. Heat (without boiling) a bottle of red wine in a pot.
2. Add about 1 kg of sugar, and dissolve into the wine.
3. Cool the mixture.
4. Soak 1-metre lengths of absorbent material (e.g. thick cord, thin rope or strips of fabric) in the mixture.
5. Around dusk, drape the wine ropes over tree branches, bushes, garden furniture or fence posts.
6. Watch for moths flocking to the ropes in the cover of dusk and darkness. They'll love the sticky sweetness and soft place to land for food!

Other Ways to Make a Moth Haven

Populations of moths are in decline, so being hospitable to these creatures is more important now than ever. But beyond sugaring and wine ropes, how else can you be a friend to moths in your garden? Here are some more simple tips:

- Moths like a mix of plants, including assorted flowers, shrubs, grasses and trees. They're not drawn only to nectar and fruit, though they do enjoy many of the same flowering plants as butterflies do.

- Moths like mulch, not rocks. If you have the choice between putting down one or the other, keep in mind that lots of wildlife will prefer mulch.

- Much like butterflies, moths will use small piles of plant trimmings, fallen leaves and twigs as shelter, and a place to lay eggs.

- Skip pesticides! They are often to blame for declining populations and moth species extinctions.

An Evening Light Trap

Want to spend an evening observing a variety of moths in your garden? Try using a light trap to draw them in.

You can use this simple at-home method for a night of peaceful moth observing or photographing. Leave out some sugar mixtures and wine ropes too, for added appeal.

The easiest light trap starts with a white sheet. Hang it in the most open part of your garden, if possible, with room on all sides (drape it over a clothesline or hang between two trees). Then shine a light or lantern onto the sheet and wait. Moths will flock to land on the sheet and sit in the glow of the light.

CHAPTER 3

Hedgehogs, Squirrels, Rabbits and Other Small Mammals

There are a multitude of mammal species that we might spot roaming around our gardens – at least if we have the right offerings on the table for them. Squirrels, rabbits, mice, hedgehogs, shrews, weasels and even foxes could all be potential visitors to your plot if you make it a welcoming space for them. There are plenty of simple projects you can try out with free supplies from nature or affordable upcycled treasures.

As with any wildlife you want to share your space with, small mammals have their good and their not-so-good moments. While watching furry little creatures ramble around your garden can be endlessly amusing and entertaining, you also need to have some patience when it comes to their less desirable habits. No matter how hard you try, you can't control wildlife; creatures will do as they please, and they will always do what they need to survive.

Inviting nature into your garden means embracing little annoyances; prepare your garden with the right projects and friendly offerings, and enjoy the brushes with nature you get to have as a result. Here you'll find tips and tricks for feeding and hosting a variety of small mammals, with an emphasis on those you're most likely to have around already.

The lesson I have thoroughly learnt,

and wish to pass on to others, is

to know the enduring happiness

that the love of a garden gives.

GERTRUDE JEKYLL

Squirrel Feeders

While squirrels are sometimes thought of as garden pests, there are plenty of good reasons to welcome them to your space; they are, for instance, sometimes responsible for planting trees and other plants, when they forget where they have hidden their stores of nuts and seeds.

Unfortunately though, squirrels will often help themselves to your bird seed and drive birds away from your garden. To avoid this issue – and let both creatures enjoy your garden amicably – place squirrel feeders in a different part of your garden from any bird feeders you might have. Then, tailor the food to each animal. It's also advisable to discourage squirrels from clambering onto your roof and nesting in your house, by keeping your squirrel feeders away from the walls of your home (and away from any trees you want them to avoid).

Squirrel Snacks

If you decide to welcome squirrels into your space, here are some treats to stock for your new visitors:

- Sunflower seeds (this is usually what they're after when digging around in your bird seed!)

- Unroasted peanuts

- Corn kernels or corn cobs

- Tree nuts

- Apples and berries

- Acorns

If you want to make your own squirrel feeder on a budget, here are three quick DIY projects you can try out:

Glass Jar Feeder

Take a large, screw-top glass jar (without the lid), strong wire and a metal fork. Place the fork on the outside of the jar (with the fork prongs poking vertically up over the lip of the jar), and wrap the wire around the two to secure the fork in place – this should act as a perch. Then, use another large piece of wire to attach the jar horizontally to a fence post or railing. Finally, fill it with peanuts and seeds for squirrels to enjoy!

Slinky Feeder

All you need for this feeder is a Slinky and a strong string or rope. Curl the spring so that it resembles a half circle, without any large gaps. Run the string through the inside curve and then hang from a tree branch, with large nuts, like whole peanuts, inside.

Peanut Butter Feeders

Spread peanut butter on a squirrel-safe surface, like a pine cone or a toilet paper tube. Roll the sticky surface in sunflower seeds or nuts, then place in a visible spot or hang from a tree branch.

Easy Hospitality

Making living space in your garden for squirrels (and other small animals) doesn't necessarily require big purchases or projects. Squirrels love to nest and hide in hollow trees, crooks between branches, and knots or holes in trunks. All you have to do to give them a living space in your yard is leave these places be! If a nice hole exists in a tree in your garden, you can rest assured that squirrels will be able to find it. You can sweeten the deal for them just by putting food near any of these ideal nesting dens, or padding them with a handful of soft leaves and grass.

Drinking Spots

Squirrels are jumpers and climbers, so will have no trouble clambering onto a bird bath you've left at eye level, or lower. Other small mammals, however, like rabbits and hedgehogs, don't have the same skills or access to those spots. If you want to make your garden especially hospitable, ensure that there's water available low to the ground every day. There are a few ways to do this: you could leave a shallow dish (with little or no lip) out on the ground or a low stump to collect rainwater; you could ensure a small divot or puddle in the ground collects water; you could even frequently refill a drinking station in a safe spot (near food offerings, but not in wide open spaces) where rabbits and hedgehogs can stop by.

Meals for Small Creatures

Ready to expand your garden meal offerings to more than just birds and squirrels? Hedgehog populations in the UK and around the world have been dwindling for decades; being hospitable to them is a great way to do your part in rebuilding their numbers and to help ensure that they'll always be around. Though you can easily buy specialized hedgehog food, you may already have some other great alternatives on hand. Hedgehogs are quite fond of tinned cat and dog food (not fish-based), which you can leave out as often as you like. You can also crush dry dog or cat treats and leave those around your fence or hedge. Don't, however, put out cows' milk, as this will make them ill.

Beyond hedgehogs, perhaps there are a few neighbourhood rabbits that you'd also like to entice to your garden with a buffet of snacks (that don't come from your hard gardening work). There are lots of foods you can leave out for rabbits, so here are just a few ideas:

- Spinach
- Herbs, including basil, parsley, coriander, dill and mint
- Unused vegetable parts, like celery leaves, broccoli stems and carrot greens
- Bok choy and chard
- Carrot tops
- Grass, hay, bark and twigs

Note that wild rabbits need to be introduced to new foods slowly, so only leave out small amounts of one vegetable at a time until you know your local rabbits are accustomed it. Never leave out avocado for rabbits as it's highly toxic for them.

Form a Hedgehog Highway

Hedgehogs spend their nights travelling – often a mile or more per night! – but their journeys become much harder when their paths are constantly blocked by impenetrable hedges and fences. If you want to be the best possible hedgehog host, talk to your neighbours and create a kind of "hedgehog highway". For this, all you and your neighbours need do is ensure that each garden provides an entry and exit for wandering hedgehogs to utilize. This may mean cutting small holes at the base of your fence or clearing out some brush to make a gap in a hedge, so do make sure your human neighbours are on board before you start making renovations for your local wildlife.

Natural Shelters

The easiest way to provide a welcoming spot for roaming hedgehogs (and other small creatures) is to make a natural shelter from outdoor materials. Start by gathering a sizeable pile of logs and branches. Fashion these into a makeshift shelter by stacking them; feel free to do so in a disorderly fashion, as this will leave gaps and spaces for animals to crawl in. Make sure there is room inside for sleepy hedgehogs, then pad the gaps with leaves and grasses. There are multiple benefits of a natural shelter: besides providing an ideal resting spot, the natural materials also attract insects, diversifying the ecosystem of your garden and providing extra snacks for hedgehogs. Besides that, they're free to make and simple for you to assemble.

Gardeners,
I think, dream
bigger dreams
than emperors.

MARY CANTWELL

Making a Hedgehog Home

While you could purchase a designated hedgehog house from a garden supplier, a little bit of extra time and a few supplies is all you need to make your own DIY shelter. Your house can be as simple as a wooden box (e.g. a wine crate) with a hedgehog-sized door cut into the front. Be sure to make the hole small enough that predators can't also slip in, as this will instantly deter your intended guests. Pad the inside with hay, leaves or soft plants. If you want to add extra security, create a small entrance tunnel with additional pieces of wood. Weigh down the house by placing large rocks on top. Complete the appeal by putting your hedgehog food and dish of water close by. And remember, hedgehogs are solitary wanderers; if you have the space, make several houses and place them in different parts of your garden.

What Not to Do

Though you now know plenty of ways to be a top-notch host to small mammals in your area, it's also wise to take notice of a few things you should avoid. Here are some points to keep in mind:

- If you use netting to support plants you're growing, they can easily become a snare for small animals walking through. If you must use netting, raise it high enough off the ground that creatures can still pass below without being snagged.

- Piles of leaves or logs and tall grasses are often claimed as resting or nesting spots for rabbits, hedgehogs, mice and others. Always check before mowing or starting a fire so that you don't mistakenly harm them.

- Skip chemicals and slug pellets. Allowing a range of creatures into your garden will turn it into a functioning mini-ecosystem. Any chemicals you add will harm wildlife.

Less Is More

In all your wildlife gardening, a simple maxim should guide you: less is more. The typical approach to outdoor spaces and gardens is to do as much as possible. While landscaping and meticulous horticulture can often look attractive, it can also deter wildlife from visiting – and flourishing in – your garden. Attend to your human outdoor living spaces in whatever way serves your purposes, but leave some room for nature to feel comfortable. Leave a part of your garden wild and free to sustain animal and plant life. A patch of tall grasses and friendly weeds is no bad thing. In fact, it's an invitation to small mammals, birds and all manner of delightful creatures that your garden is a safe place to stop by and make themselves a home.

CHAPTER 4

Amphibians
and
Reptiles

Depending upon the climate and environment of your outdoor space, attracting reptiles and amphibians could be a real possibility. Herptiles (more commonly shortened to *herps*) are the broad category of cold-blooded creatures that slither and slide about – frogs, toads, snakes, lizards, newts and slow-worms are all examples of these. These creatures are found in a range of climates and will flourish in most places where they have access to sun and water. Fully committing to life as an at-home herpetologist can be costly and time-consuming, but a few simple, affordable projects can have you well on your way to harbouring these animals in your own garden.

Unfortunately, many amphibian and reptile populations around the planet are steadily decreasing. Typically, this is the result of a loss of habitat. However, by striving to provide habitats in your own garden – however small or simple – you will be creating a space for these creatures to survive and thrive. If we all did a little bit to help, the result could be overwhelming. And, of course, any animals and wildlife you attract to your garden will cause a chain reaction; more and more creatures will follow them, turning your outdoor space into a wildlife haven.

We may think we are

nurturing our garden,

but of course, it's our garden

that is really nurturing us.

JENNY UGLOW

Hospitable Gardens

Attracting reptiles and amphibians is all about making wet spaces and avoiding the introduction of deterrents or predators. However, before you start adding ponds or water features, make sure your garden is accessible. As with hedgehogs, if small frogs, lizards and other herptiles can't get into your plot, they certainly won't show up in your pond. Providing that highway or wildlife corridor is very important, so (if you can) leave gaps in fences or hedges to let the nature in. Another easy way to be hospitable is to vary vegetation heights. Dense, low vegetation gives ample space for hiding out, and open, sunny spots let cold-blooded animals rest comfortably to warm up when they feel safe.

Wet Habitats

If you're particularly keen to bring a troupe of frogs, toads and newts into your garden, the best thing you can do is establish a pond. While you may not have the space for such a feature (we'll explore smaller-scale options later), it is one of the surest methods to attract reptiles and amphibians. Even a small pond will be a great draw. If you're looking to add a pond to your garden, consider the following tips:

- Keep the pond and surrounding spaces as natural as possible. Simplicity is key.

- Choose a particularly sunny part of your garden.

- Leave shallow space around the pond for animals to come in and out. (Not paving slabs; these get too hot.)

- Stock with a variety of native plants.

Not enough space for a full pond? You could opt for a small bog garden instead. Rather than standing water, a bog garden is a patch of waterlogged plants and soil that has similar conditions to a real bog. It allows you to grow marshy plants and to attract small amphibians. To make one:

1. Choose a sunny spot and dig a hole, about 50–100 cm deep.
2. Cover the hole with a butyl (or rubber) liner and pierce several times so it can drain.
3. Place soil on top of the liner, to fill the hole (then trim away the liner around the sides). Include a small amount of compost.
4. Add moisture-seeking plants, then water thoroughly to get it started.

Pond Safety

If you're adding a pond to your garden, you should always make sure it's safe: both for the animals you're making it for and the people you live with. If you have children, you may wish to put a barrier around the pond. As far as avoiding hazards for the amphibians and reptiles making use of your pond, here are some tips:

- If using tap water to fill your pond, let it stand for a couple of days before you introduce any creatures to it.

- Don't clean your pond or add chemicals, and keep pesticides out of your garden. By allowing your pond to mimic nature, you'll have better luck attracting wildlife.

- If possible, have a section of the pond that reaches at least a 60-cm depth. Some frogs will spend the winter at the bottom of a pond, so this will make for a habitable spot for them.

Frog and Toad Houses

A simple DIY frog or toad house can be made using either a plastic or a clay pot. Though using a plastic pot is cheaper and simpler, it's worth bearing in mind that it will also get hotter in the sun. If using a plastic pot, cut an entrance hole in what would normally be the top lip, and push the pot firmly into the soil (top down) – ensuring that the cut opening leaves enough room for a frog or toad to amble inside.

If you prefer to use a clay or ceramic pot, you can try one of two methods. You can either make a small ring of rocks to place the clay pot on top of and then remove a few stones on one side to create an entryway, or you could turn the pot so it's positioned horizontally along the ground, then dig the opening into the soil.

Alternative Habitats

Whether you're confined by the space you have, your pre-existing landscaping or your budget and materials, the lack of a pond doesn't mean you won't be able to welcome amphibians and other reptiles into your garden. There are plenty of other DIY habitats you can provide for local frogs, toads, slow-worms and newts.

Compost Pile

While enriching soil and housing beneficial fungi and bacteria, compost piles often become homes to slugs and small insects. This, in turn, will entice creatures like slow-worms and frogs to stop by to forage for food or take shelter.

Rockeries or Log Heaps

Both piles of stones or piles of logs and sticks are attractive to amphibians and lizards. The small spaces make ideal hiding spots for them, and the nooks and crannies attract tasty insects for them to snack on.

Mini-Pond

Rather than digging in a pond, you could opt to provide a mini-pond environment for your local wildlife. While it's unlikely to become a nesting or hatching spot, it could still be a destination for amphibians and reptiles to lounge, cool off or snack. Any large tub or container can become a DIY pond; simply place it in a sunny spot, fill it with water and add a variety of water-loving plants. Note that ponds placed at ground level can allow easy access to predators.

Natural Pest Control

If you're struggling with some uninvited guests in your garden, aim for natural solutions. Remember that by welcoming wildlife into your garden, you are allowing your space to become a small ecosystem with a food chain. Those creatures which you may consider "pests" are likely to be feeding the birds, mammals, frogs and other creatures that you've invited into your garden. However, if you do find slugs (or other creatures) are snacking on your plants or damaging your garden in some way, there are some natural ways to deter them:

- Sprinkle ground coffee around the plants which slugs are most attracted to. This will deter them, as well as benefit your plants when the grounds decompose.

- Leave crushed eggshells around plants at risk. The sharp edges will often deter slugs from climbing over and onto your plants.

- Try spraying plants with soapy water (nature-safe soap), pepper spray or neem oil spray.

Beware of Predators

Small reptiles and amphibians have plenty of natural predators. While this is part of the food chain, you can also do your part by protecting them from your own pets. Wandering cats can often mean the demise of a frog or toad. If you have cats that spend time outside, monitor them while they roam, or keep them in areas away from your pond or frog-sheltering spots. Ponds naturally offer shelter for submerged animals, but providing other places for cover (like the suggestions given earlier) is also an easy way to give small creatures an escape route if predators do come around.

Reptiles and amphibians...
can be lethally fast, spectacularly
beautiful, surprisingly affectionate
and very sophisticated.

DAVID ATTENBOROUGH

Wintering

As cold-blooded creatures, amphibians and reptiles are sensitive to extreme temperatures. In the winter, many species will hibernate, either settling into the mud at the bottom of a pond, or finding a safe, sheltered bit of ground to dig into and hide out. If you've added a pond to your garden, it's likely to become the winter home to some small creatures. If you don't have the space for a pond, or prefer another route, you can always create a safe hibernation refuge. Amphibians feel safest in small spaces, so before temperatures drop too low, stack a small pile of wood and leaf litter in part of your garden. Pad the empty spaces with some wood chips, plant trimmings or loose soil. The log pile will attract tasty bugs for reptiles and amphibians to snack on, and the soil will become a great burrowing spot during the coldest parts of the year.

Take Cover

Another way to be hospitable to frogs, toads, newts or slow-worms is to provide plenty of options where they can take shelter and hide from the elements or predators. As mentioned previously, small creatures face plenty of risks, and their natural instinct is often to hide and wait for safety. Plenty of household supplies can become herp hideouts.

Wooden Sanctuary

All you need for this is a large thin sheet of wood which you can find fairly easily at a hardware store. Lay one or more sheets down in the spots you think might contain herptiles. Sunny corners or anywhere near a wooded area or pond will be a safe bet.

And that's it! If you ever want to take a look for creatures hiding underneath, raise the board slowly to get a glimpse of lizards or other resting stowaways.

PVC Hideouts

Have some empty PVC tubing with no purpose? While most of these projects have been suggested with ground-dwelling animals in mind, there are also plenty of frog and lizard species that spend their time in the trees. If you have a wooded area in your garden, or even a couple of trees, this method may be effective. To hang, drill holes at one end of your pipe and thread rope or string through, then tie around tree branches or trunks. Ensure the pipe hangs horizontally, so that creatures can live inside. Come back later to see who is making use of your DIY hideaway!

CHAPTER 5

Bees
and
Wasps

As one of nature's most necessary workers, bees are critical to human life as we know it. Among other things, they pollinate our plants – in fact, it's believed that about a third of the food consumed in the world every day relies on bee pollination to grow. Though there are other pollinating insects and small animals, honeybees make up the largest percentage by far. They're even responsible for the growth of non-food crops (plants we use for clothing, medicine, cosmetics and cleaning supplies). We rely on them, but their numbers have started to decrease rapidly in recent years (largely due to the human effect on the environment). Knowing all of this, and aiming to be the best steward to nature and wildlife that you can be, creating bee-friendly spaces and planting for bees is an easy thing you can do to make a big difference. If every home gardener is a good host to bees, we'll collectively be able to support their populations and keep them from being lost for good.

Wasps, too, have their role in our ecosystem, some acting as pest predators, others working hard as pollinators. As you grow flowers and fruit plants in your garden, you may start to host bees and wasps unintentionally; this is a start. But there are so many ways to dig deeper and do more for these flying, pollinating neighbours. The following pages provide a variety of planting recommendations and natural projects to attract bees and wasps into your garden.

Bees do have a smell, you know, and if they don't, they should, for their feet are dusted with spices from a million flowers.

RAY BRADBURY

Happy Bee Spaces

While you might have been led to believe all bees live in grandiose beehives as a large team, there are some bees who live and nest alone, going about life without a huge hive family. These solitary bees still look for places to nest and take shelter. Most often, they prefer loose, crumbly soil where they can dig in and feel safe. If you have an area of your garden available, leave loose soil behind there – especially if it's a sunny spot – as this will be appealing to passing bees. If you can, position this area of crumbly soil near plants and flowers that bees love.

Flowers of Choice

If you're looking to attract bees to your garden, planting lots of flowers is a great start. However, it's critical to remember that all plants have different flowering seasons. While some may bloom fully in early spring and then die off, others may be growing all season but only flower in the summer or autumn. In order to be the best host you can to your local bees, plant a variety of flowers, herbs and shrubs with a mixture of blooming seasons. Here are a few you could start with.

Spring Blooming:

- Apple or crab apple
- Calendula
- Crocus
- Flowering kale
- Lungwort
- Marjoram
- Pussy willow
- Wild lilac

Summer Blooming:

- Bee balm
- Chive
- Foxglove
- Hawthorn
- Lavender
- Phacelia
- Snapdragon
- Strawberry

Autumn and Winter Blooming:

- Abelia
- Honeysuckle
- Perennial wallflower
- Rosemary
- Sedum
- Snowdrop
- Witch hazel
- Zinnia

Building Your Blooms

Beyond having blooms throughout the seasons, varying your bee-friendly shrubs and flowers has several other key benefits. There are a variety of bee species, each with different lengths of tongues, designed and adapted for particular styles of flowers. That means without some variety of flower types, some of your bee visitors may fly off still hungry, unlikely to return. To keep your bees coming back, mix and match flower types. In general, you should avoid flowers that are double or multi-petalled, as these often lack nectar or pollen – meaning they won't entice pollinators like bees or butterflies. Colour is also important. Bees are particularly drawn to yellow, purple and blue blooms. Last but not least, do your research before planting. Stick to native plants for your area only, as these are the ones that nearby creatures will be more attracted to – and they won't negatively impact your garden ecosystem.

Welcome Wildflowers

If you've ever walked through a field of wildflowers, you've likely seen a multitude of bees buzzing from plant to plant. Wildflowers, especially in great quantities, will draw bees more than anything else. Since it's a good idea to leave a bit of your garden wild for animals, letting that spot fill up with wildflowers is one great route you can take. Wildflower seeds can be purchased from any garden shop – often in assortments. A small dedicated wildflower garden will require no maintenance from you, but will offer a great reward. Once they're blooming, you're sure to see bees come to join them (plus butterflies, moths and other insects!).

Poke Holes for Bees

If you have a large garden with a wet space like a pond – or are fortunate enough to live on a riverbank or other watery spot – consider providing these individual nesting spots for your local bees. All you need to do is take a pencil and then poke several spaced-out holes into some sandy soil with it. Be careful to take note of where you make holes; if they end up becoming bee nesting spots (as you intended), you won't want to later start digging up or trampling around those areas.

If you have a lack of space or sandy soil in your garden, you can also try to safely use this method in your neighbourhood to support the world's dwindling bee population. If you know of any fitting place nearby – a secluded riverbank, lakefront, pond or creek that isn't crowded with people – take your pencil and leave some nesting holes there.

Sugar Water

If you ever come across an exhausted or struggling bee in your garden or home and want to help, there's a quick solution that might help to revive it. Mix two tablespoons of white sugar with a tablespoon of water and leave it on a spoon next to the bee. This should offer them enough energy to return to their hive or home.

However, try to avoid leaving sugar water outside at all times. Despite being a quick fix, it can be detrimental to their hive. Sugar water is lacking in nutritional value, and bees may ignore flower and pollen in favour of it.

Build a Bee Bungalow

A bee bungalow is a DIY project you can make in a day and that can become a home for your local solitary bees for their entire life cycle. Free-roaming bees need places to live and nest, and if they relocate to your garden, they're going to spend their days pollinating your plants. It's a win for both of you.

To make one, you'll need:

- An empty wooden box with an open front and sloped roof. You can make one yourself with scrap wood or repurpose a bird house or other wooden box.

- Hollow plant stems and/or pieces of wood (which will fit inside your box).

- A drill or means of making holes.

Prop the box (open side facing out) in a sunny spot in your garden, at least a metre from the ground. This could be along a fence or treeline, near your flowering plants, or away from common seating areas. Stack wood and stems inside the box, creating rows. Drill holes into the outward facing sides of the wood, being careful to make the cavities smooth and splinter-free.* Diameters should be between 2 and 10 mm, though this varies based on the particular bee species you're looking to accommodate. If you want to be extra prepared, do some research into which bee species live in your area. Step back and watch as bees make your bungalow a home.

*You may wish to purchase ready-made bee tubes, if you lack the time or resources for this step.

The hum of bees
is the voice of
the garden.

ELIZABETH LAWRENCE

Becoming a Wasp Ally

Wasps are close relatives of bees, though are more commonly considered as pests. While you may be fearful of wasps, there are plenty of reasons to reconsider your feelings toward them. Most wasp species are only aggressive when they feel threatened (yellow jackets are the ones to watch out for), so the first place to start is developing a calm and cautious attitude when they're around.

Wasps, like bees, are pollinators. Often considered generalist pollinators, they help to carry pollen between many types of flowers and plants. They're also critical to the pollination of fig trees. Figs and wasps couldn't exist without each other, as they're each instrumental to the other's life cycle. And beyond that, wasps are predators of many common "pest" insects, making them an especially valuable part of any garden ecosystem. So, don't turn your back on wasps too quickly; they're better to have around than you may think!

Coexisting with Bees and Wasps

Bees and wasps are a hugely important part of our ecosystems and even our economies. Protecting their populations is as simple as learning to coexist peacefully. Here are a few helpful tips:

- Teach kids from a young age not to fear bees and wasps. If you react with fear when you catch sight of a bee, the children around you will likely copy your behaviour.

- Be calm and try not to make jerky or aggressive movements.

- Pay attention to the bee or wasp you're looking at. Out of thousands of species, only a few are actively aggressive. Unless you know you have a yellow jacket around, you likely shouldn't worry.

- Stay away from swarms on the move. They're likely to be more dangerous – mainly because you're outnumbered.

If You Find a Wasp Nest

It's possible that at some point in your life, part of your home, garage or garden will become host to a beehive or wasp nest. While this may be a cause for alarm, you needn't panic. The hive has likely been there longer than you realize, without hurting you.

If you do end up hosting a nest outdoors (away from frequent human interaction), this could be a positive thing. Bees and wasps are amazing to have around your garden, so try leaving them to live peacefully and pollinate your plants.

However, if the nest is indoors, it's always safest and wisest to call a professional to help you. Look for someone who uses methods that don't kill the creatures and who will move and re-establish the nest in a safer spot. Always avoid destroying a nest if you're able to; you wouldn't want someone coming by to damage your home either.

CHAPTER 6

Dragonflies
and
Damselflies

Dragonflies (and their close relatives, damselflies) are fascinating to observe. They flit quickly from place to place, stopping to hover in mid-air or pause on a sun-soaked leaf. They're iridescent, painted in shining shades of green, blue, purple, gold and even occasionally a shocking pop of red or orange. Up close, the intricacies of a dragonfly body or wing are remarkable. But it isn't only their beauty we should prize them for; these water-loving insects are also a valuable asset in your garden.

Dragonflies and damselflies spend most of their lives underwater, and only have a short lifespan once they're flying about for you to see. With better vision than humans, these arthropods are the foremost predators in the insect world. They zip around snacking on mosquitoes, gnats, cicadas and other bugs; dragonflies have a prey catch rate of 95 per cent, and just one dragonfly can eat upwards of 100 mosquitoes per day.

So, if not just for their enchanting looks alone, try to make a home for dragonflies and damselflies in your garden. Your garden food chain will be vastly improved, and you might just be able to spend summer nights outside without falling prey to a swarm of mosquitoes. All thanks to a few hungry bugs.

One touch of nature

makes the whole world kin.

WILLIAM SHAKESPEARE

Dragonfly Basics

Not unlike amphibians and some reptiles, dragonflies and damselflies are water-loving creatures. These insects need water for laying eggs, lounging around and for attracting their prey. As a result, many of the tips recommended for making space for frogs, toads, slow-worms and newts also work for drawing dragonflies. Be aware, though: if you have a pond and want it to double as a home for amphibians and for aquatic insects, the insects will become prey. In a pond with frogs or fish, dragonfly eggs laid in the water will be eaten by your other wildlife inhabitants. So, choosing which species you prefer to attract with a pond is an important step you should keep in mind when undertaking any wildlife garden renovation.

Small-Scale Water Features

Not all gardens have the space for an in-ground pond. If you have a balcony, courtyard, patio or small yard, don't worry. You can still provide your own pretty and affordable version of an aquatic insect paradise.

You can create your own mini-pond for dragonflies with a wide variety of containers. Many pre-built fountains with trickling or bubbling features are a perfect choice. As long as it has a reasonable depth, it should do the trick. You can also build your own water feature from items you salvage or upcycle. For instance, a large steel bucket or plastic tub could easily be transformed into a pond-like oasis. For any attempt at a dragonfly-friendly pond, keep these requirements in mind:

- Shallow edges make it easier for insects to lay eggs.

- Parts of your pond should be at least half a metre deep.

- Include lots of plants.

- Find a spot with about 30 per cent shade, if possible.

- Don't introduce fish! If needed, have two different water spaces, one for fish and frogs and a smaller one for aquatic insects.

- If you don't have a moving water element in your mini-pond, it will likely become a breeding spot for mosquitoes, as well as dragonflies. While dragonflies and damselflies are predators to mosquitoes, you probably still don't want to invite more of them. Add a bubbling or moving component to your pond to prevent stagnant water.

Floating Plants

A key to any great dragonfly water feature is the inclusion of some native plants. Selecting plants that float make for both a visually attractive component for your pond and a practical spot for insects to lay their eggs or rest. Some of the best pond plants are those considered free-floating. They're very low maintenance, as all you should need to do is place them onto the top and let them spread their roots into the water and grow as they please. Some common floating plants include:

- Water lily

- Duckweed

- Water soldier

- Fanwort

- Water hyacinth

Underwater and Swampy Plants

Aside from free-floating, surface-dwelling plants, there are a variety of other beneficial plant types you could consider. Submerged vegetation (plants living entirely below the water) provide cover for eggs and young dragonflies, while oxygenating the water. Emergent vegetation (those rooted at the bottom of the water with portions growing up and out of the water) give nymphs a path from the water to the land. Shoreline plants that grow along the wet edges of a pond provide spots for grown dragonflies and damselflies to perch, rest or hide.

Some popular submerged plants include hornwort, mermaid plant, waterweed and arrowhead. Emergent plants include cattails, bulrush, reeds, lotus and horsetail. As for shoreline plants, any plant native to your area that thrives in very wet, swampy soil will be perfect.

A Waterlily Project

A waterlily is an attractive plant for dragonflies and humans alike. They thrive in small pond environments and they provide a perfect surface for egg-laying or resting predator insects. Better yet, for the garden-owner, they're beautiful and peaceful additions to a water feature, easy to plant, and provide nice shade to keep algae in check. However, waterlilies can also be a bit aggressive – they can take over a pond and smother other plants if left unchecked – so this project keeps them a bit more contained and under control without losing any of the aforementioned benefits.

1. Select an aquatic planting container and line it with hessian.
2. Fill about three-quarters full of aquatic soil (preferably one dense with organic material).
3. Plant the waterlily in the soil.
4. Backfill the container with more soil, leaving the top part of the plant exposed.
5. Add some rocks or gravel to the top, to keep the soil from floating out.
6. Slowly place the lily container into your pond, about half a metre deep at most.
7. As the weather gets colder, move the lily in its container more toward the centre of the pond to keep it from freezing.

Rock Piles

Dragonflies and damselflies, like all insects, are cold-blooded. Though it's possible for them to overheat, it's more common for them to be cold and need to find a warm place to get some sun and regulate their body temperature. Cold dragonflies actually have a strategy (similar to shivering) called wing-whirring, that helps them warm up. You're likely to see them doing this on a cold morning, before being able to fly. If you want your garden to help chilly dragonflies warm up, leave a small rock pile in a sunny spot. The darker the rocks, the more heat they'll trap. As with reptiles and amphibians, these warm rock hangouts will be perfect spots for dragonflies and damselflies to warm up their wings.

Another Win for Pollinator Plants

As discussed, pollinating plants are an asset to any garden. They attract bees, butterflies and even some small animals. By including a variety of plants – many of them pollinating – you're priming your garden to become a mini-ecosystem of its own, allowing all types of wildlife to flourish and thrive.

When it comes to pollinating plants, dragonflies and damselflies will also benefit, though in other ways. Since many smaller pollinator insects will be attracted to the plants, dragonflies will be quick to show, looking to snack on the variety of insects in your garden. A win for both of you!

To nurture a garden is to feed not just the body, but the soul.

ALFRED AUSTIN

Dragonflies vs Damselflies

Now that you've seen their benefits and how exactly to keep them around, you may still be wondering about the differences between dragonflies and damselflies, and how to spot them. They are very closely related, but there are a few key differences you can rely on to determine which one is visiting your garden. The biggest difference is in the wings. Both have two sets of wings, but the back set of wings on dragonflies are larger than the front set, and for damselflies all wings are the same size and shape. You can also wait to see one at rest – dragonfly wings stay perpendicular to the body, while damselfly wings fold up. But, ultimately, both species are equally beautiful and valuable in your garden ecosystem.

CHAPTER 7

Minibeasts

Making nature a part of your space means taking "the bad" with the good. There will always be creatures you may consider to be pests but those "pests" are part of the larger fabric of our ecosystem. Creatures you may consider undesirable are often a source of food for the insects you hope to attract. Likewise, the insects you attract might then serve as a foodstuff for birds and reptiles and countless other members of the animal kingdom. Small organisms also do lots of work "behind the scenes"; while some creatures pollinate your plants, others may compost your food waste.

Minibeasts – as we'll call them in this chapter – make up a huge and diverse part of the natural world. It's inevitable that many of these creatures will be a part of your garden, whether invited or not. Flies, beetles, spiders, ants and even some little crawlers you can't see with the naked eye, all come under this category. On every part of our planet, insect species (both those we've identified and those we have yet to discover) roam and coexist with us – often providing their services without us even knowing it. This is your reminder to make room for the minibeasts in your garden. Be kind to *all* of nature and nature will be kind to you.

Some people talk to animals.

Not many listen though.

That's the problem.

A. A. MILNE

Your Quick Guide to Minibeasts

You probably already know a few of the insects who reside in your garden. Though lots of bugs tend to be unpopular, there are millions of insects providing important services to our plants and who make up a critical part of our ecosystem. Here are just a few helpful species:

- Springtails (tiny water-loving creatures, common across the UK)

- Soldier beetles (these beetles have narrow, rectangular bodies and tend to be red or orange in colour)

- Ground beetles (ground-dwelling and largely predatory, these beetles are very common in the UK)

- Lacewings (taking their name from their delicate transparent wings, these bugs are aphid predators)

- Ladybirds (common, colourful bugs, with a penchant for plants)

- Spiders (largely predatory, spiders are great for helping with "pest" control)

- Centipedes (long, predatory insects which will also help control "pest" populations in your garden)

Build a Minibeast Hotel

Building an all-purpose bug hotel is not too dissimilar to building a solitary bee hotel – you could even create one hotel that works for both types of insects. While bees will appreciate logs and sticks with holes drilled into them, a variety of other materials can be combined to make an insect haven for a wide range of minibeasts. Though, by making a bug hotel, you do run the risk of attracting some "less desirable" insects, you will also be inviting predator bugs, who will likely eat them and strengthen your garden ecosystem.

A minibeast hotel is a highly customizable DIY project. Here are some of the many components you can include:

- Sticks, branches and logs
- Pieces of wood or wood shavings
- Dry leaves or grasses
- Unused pots, pallets or buckets
- Bamboo pieces
- Stems, seed heads and other plant materials
- Pine cones
- Lichen or moss

To put your hotel together, collect any of the supplies you'd like from the list (you might even have some of the items in your garden already). Then, choose a container: it could be an old wooden pallet with openings, a large pot laid on its side, or a wooden crate similar to the bee hotel (page 83). With one or both sides open for insects to crawl in, stack up layers of your natural items. Piles of sticks, patches of grass and leaves, and even plants or pine cones will all create different spaces for different insects to make their home.

Be Less Tidy

If there are overgrown parts of your garden – or if you have dead plants or trees lying unused – don't worry! Consider leaving that dead log or pile of plant trimmings for a while. As long as they're not doing any harm or getting in your way, they can provide shelter for all manner of wildlife, including helpful minibeasts! Insects will enjoy eating away at your plant matter, and in doing so they'll essentially be cleaning up for you. At the same time, the great food source will keep those minibeasts around with their populations growing. Why not let nature do the work, while you help it thrive?

Log Piles: Minibeast Hideaways

A log pile is another practical inclusion in your garden. Beyond appealing to amphibians, reptiles and mammals, log piles (even small ones) are havens for minibeasts too – even those too tiny for you to see. A pile of logs and branches left in nature will become a quick host to microorganisms, which then begin breaking down the wood and preying on "pests". Larger insects, too, will be drawn to the pile, to prey on smaller creatures. From there, the log pile will have enough insects for frogs, toads, slow-worms, newts, hedgehogs or mice to take note (since all of these creatures enjoy snacking on bugs). Before you know it, one small log pile placed in your garden will be sustaining its own tiny food chain, and you'll become a host to wildlife with minimal effort!

Warm Spots

All insects are ectothermic, or cold-blooded; this means they seek sunny places and warm spots to survive. Some of these warm places might not take any work to create, while others can easily be created and added to your garden.

Here are three warm garden spots where minibeasts will thrive:

Sunny Vegetable Patches

If your garden area gets lots of direct sun, it will become an insect resting spot, and you're likely to see some visitors crawling or flying around.

Concrete Walkways

If you have a path into your garden or around your house that gets sun, insects will seek these warm spots out.

Rockeries

Simply leaving a nice pile of rocks in a sunny spot is helpful for all manner of insects, and even larger creatures too.

Select Flowers Wisely

Mouth parts vary greatly in size and shape between insect species, since each insect has a mouth specifically suited for particular types of plants and flowers. A butterfly proboscis is designed to fit perfectly into long, tubular flowers. For insects and minibeasts with very small mouth parts, flowers with open, flat petals will be most accessible. Try not to worry too much about insects eating your flowers; most of the bugs that do this are so small they won't be able to make much of a dent. Some great flower options for your garden include asters, daisies, carrot flowers, dahlias and sunflowers.

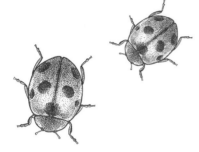

Planting to Attract Minibeasts

While it's impossible to cater specifically to every type of minibeast with the plants you choose for your garden, you might want to give yourself a better shot at appealing to particular creatures. Have a need for predator insects? Looking to entice some extra pollinators, like beetles and flies? Just looking to enhance your garden's biodiversity and help its ecosystem thrive? Whatever your goal, there are minibeasts to help you. Here's a list of some helpful minibeasts likely to be living in your garden, and the plants they enjoy.

Lacewings

These all-purpose pest-killers love sweet alyssum, cosmos, dill and caraway.

Hoverflies

Hoverfly larvae feast on aphids, and the adults love flowers like marsh marigolds, fennel, myrtle and single-flowered dahlias.

Damsel Bugs

Highly effective hunters, these petite bugs enjoy alfalfa.

Ladybirds

Also commonly known as ladybugs, these recognizable creatures are friends to gardeners, and will be attracted to coreopsis, dill and fennel.

Pirate Bugs

These hardcore hunters with a feisty name will be attracted to yarrow, alfalfa and daisies.

Ground Beetles

Their larvae have insane appetites and dine voraciously on caterpillars. Planting rhubarb or lovage, mixed with other plants, will help them thrive.

My garden
is my most
beautiful
masterpiece.

CLAUDE MONET

The Benefits of Composting

It's likely that you've heard people preach the many benefits of composting. Among other things, it adds nutrients to soil, helps to recycle kitchen waste, creates beneficial bacteria, suppresses plant diseases, reduces landfill and is great for the environment. Compost bins also provide the ideal living conditions for a range of helpful minibeasts. After all, these warm, natural environments provide creatures with a range of places to hide and a great deal of plant matter to eat. Moreover, plenty of tiny insects and bacteria thrive in these conditions, which provides a source of food to larger creatures.

What Is a Wormery?

Not unlike a traditional composting set-up, a wormery involves collecting plant scraps and kitchen waste to be slowly converted into nutrient-dense soil. The difference is that in a wormery, specialized composting worms are added to speed up the process. A wormery can be just as space-efficient and practical as your average compost bin and will also attract tiny (and often unseen) minibeasts. A home-made wormery works quite simply: worms live in a top compartment, where you also place plant trimmings and kitchen scraps (such as vegetable peelings, tea leaves, coffee grounds and bread). Worms quickly and effectively break down the plant matter and convert it into compost, ready to go into the garden. A lower compartment also collects liquid, which can be poured over plants. They're quick and painless to make and will be an asset not only to your plants, but to your wildlife environment as well.

Making Your Own Wormery

Intrigued by the concept of a wormery and ready to try one out at home? It's fairly simple to make your own DIY version. You'll need:

- Two (or three) boxes or trays that fit together – one with a lid

- Composting worms (look for specialized composting worms, which eat much faster and can be ordered online)

- A drill

- Compostable materials

For a simple, two-layer wormery, drill half-centimetre holes every 5 cm or so across the bottom of the top box, with the lower box then stacked below it to catch run-off water. Add a thin layer of bedding material to the top box, followed by your worms and a layer of kitchen waste. Add a single row of small holes in the lid of the top box, to provide airflow. Wait a week, then add more kitchen waste as you have it. Keep the wormery somewhere shady, ensuring that it doesn't get too warm or cold.

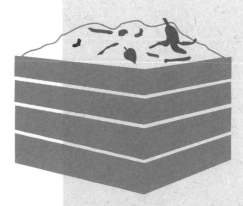

CHAPTER 8

*Fungi
and Moss*

Though they might sound rather less exciting than some of the wildlife discussed so far, fungi and mosses both have significant roles to play in our garden ecosystem. As masters of decomposition, premium soil purifiers and environmental allies, moss and fungi operate across the globe, and do a lot of good in the process.

You're probably most accustomed to seeing moss or fungi outside of your own garden. They pop up in all sorts of ecosystems and climates, clinging to the sides of trees or fallen logs, forming clusters in forests or grassy fields. Among the more than 144,000 fungi species identified on Earth, you'll find everything from tasty edible mushrooms to poisonous toadstools to microscopic colonies and endlessly spreading underground fungi. In fact, the largest organism on Earth is not an animal, but a massive single colony of *Armillaria ostoyae* above and below the ground in Oregon (US), spanning a massive 3.8 km. Mosses, too, are incredibly varied, with over 10,000 unique species around the world, all of them with their own preferences and patterns.

Both mosses and fungi are among the most intricate and stunning organisms to observe. Though they won't fly about your yard, or crawl into a little den, they can still be fascinating to look at and foster in your garden. Besides their host of benefits for the plants and animals making your garden their home – and their positive environmental effects – these charming natural wonders are fun to cultivate and improve the look of your outdoor space all at once.

By the meditation of a thousand

little mosses and fungi,

the most unsightly objects

become radiant of beauty.

HENRY DAVID THOREAU

Tread Carefully

While fungi can be great sources of food, learning to properly identify wild mushrooms is a long and tricky process. With over 10,000 mushroom species across the world, many of them look very similar. Unfortunately, many species of fungi are poisonous, causing nausea, illness or even death, if ingested by humans or pets. Expert mycologists can spot a poisonous mushroom versus an edible one with little trouble. But, as most of us are likely to be novice mushroom hunters and growers, it's critical that you take extreme caution before ever eating a mushroom you forage or grow. The focus in this chapter is instead to look at the advantages of fungi in the garden space, and ways to grow or cultivate them to benefit your flourishing outdoor ecosystem.

Garden and Ecosystem Benefits of Fungi

Some people associate the word fungi with edible mushrooms growing on the sides of logs or in the garden; others associate it with mould – the fungus known for invading homes, causing illness or killing off plants. Though mould does come under the category of fungi, most of its relatives are much less concerning. In fact, most mushrooms and fungi provide great benefits to ecosystems and other plants. Here are a few of the positive benefits you reap having fungi in your garden or outdoor area:

- Mushrooms indicate healthy soil. If you see mould growing on a plant, it is a cause for concern, but seeing mushrooms popping up (even on indoor plants) is generally a good thing. The presence of mushrooms means the soil is rich and healthy, and that some lingering spores have had the chance to grow. This is a good sign for plants.

- Mushrooms decompose organic material into food for plants and soil. This includes tree stumps, logs, dead leaves and wood chips.

- Below the ground, mushrooms have a large network of root filaments, which form a relationship with plant roots. This provides plant roots with more surface area and more carbon, as well as better disease resistance and drought survival.

- They are a hugely important part of the food chain and help diverse ecosystems thrive. Every food chain needs decomposition to function, as it allows the energy of dead organisms to be recycled and used again.

Invest in Wood Chips

If you already grow plants in your garden, you may be familiar with the many advantages of wood chips or mulch. There are a wide variety of garden benefits from wood chips: not only do they retain moisture in soil (keeping plants healthier), but they also break down naturally, control weeds and pests and prevent soil from eroding away. The surprise benefit? They also help many mushroom species thrive.

For the most success, put wood chips down in a shady, cool area near the base of a tree (or several trees). Ideally, soak the wood chips for several days or more before laying them down, to help kill off bacteria. From there, you can leave them and let nature take its course. Or, to be certain you'll have mushrooms, sprinkle a bag of fungi spawn across it. Be sure to choose a species that is safe and non-toxic.

Establishing Fungi on Logs

If you've ever walked through a shady forest, you've likely noticed different mushrooms and fungi growing on logs, trees and in brush piles near them. These are the types of environments where wild mushrooms flourish. If you want to plant some log-dwelling mushrooms of your own, there are ways to do so. The simplest route is to find and purchase mushroom dowels (which can be found online). These are wooden implements impregnated with particular fungi spawn, which can be inserted into wood. For the best chance of success, insert them between autumn and spring.

Choose your log or stump (hard woods are best) and allow it to dry out. Drill holes in the log about 15 cm apart. Insert the dowels and lightly hammer them in so they don't protrude from the log. Position the log in a shady part of your garden, or beneath some dense shrubbery if needed.

Growing and Harvesting Edible Mushrooms

Though, as mentioned before, it's very dangerous to collect and eat unknown mushrooms without an expert involved, growing safe mushrooms in your garden is an activity you may decide to undertake. If you decide to pursue growing your own mushrooms, always do a lot of research before planting anything, and don't eat the mushrooms you grow. There's always the risk that extra spores will be growing nearby and you could end up eating a mushroom other than the one you intended (unless grown indoors). Mushrooms are beautiful and fascinating to grow and observe, making for a great gardening project that also supports your ecosystem. Mushrooms like dark, cool, damp growing environments, meaning some will even thrive in your basement.

If you do decide to become a mushroom cultivator, here are a few of the varieties you might have most luck with as a beginner:

Pearl or King Oyster

These popular types of mushrooms are members of the large family of oyster mushrooms, and are two of the most popular and easy to grow. These are adaptable species that will grow in many climates and materials.

Shiitake

This popular variety is best grown on logs, so the log-growing method comes into play here. The growing time is longer on these, about six to 12 months, so planting once a year after harvest is a good practice with shiitakes.

Benefits of Moss

Ever thought about replacing your grass with moss? It may sound strange, but that's just because we're accustomed to believing that gardens have to have grass and landscaping. However, moss is slowly starting to gain some recognition as a more practical, sustainable option for outdoor spaces. Even if you don't cover your whole plot, growing some moss in your garden has countless benefits. Here are a few:

- Affordable and easy to plant
- Requires minimal water, especially compared to grass, so could cut your water bill while being sustainable
- Grows and thrives year-round
- Attractive and landscaping-friendly
- Little maintenance required, and no fertilizers or chemicals used for growing
- Natural weed blockers
- Hearty and can bounce back on its own; can even absorb some toxins from the soil
- Can purify unsafe water sources and reduce soil erosion

What About Lichen?

Lichen and moss are often connected by visual association. We think of them as similar – perhaps even the same – as both grow across trees and fallen logs. But what are the real differences?

Essentially, mosses are plants and lichens are not. Lichen is actually a fungus, but a peculiar type that can't exist without a symbiotic relationship with algae or bacteria. Lichen varies in colour and appearance; some are flaky or scaly while others form a thin layer across a surface. Most reproduce like fungi (through spores).

Lichen isn't harmful to plants and can even supply some helpful nutrients. Besides that, they provide shelter for other tiny organisms like bacteria and insects. They also act as food for some animals, or material for their shelters. So, if you see some lichen in your garden, welcome it – it's just another great part of your mini wildlife-sustaining ecosystem.

I don't like formal gardens. I like wild nature. It's just the wilderness instinct in me, I guess.

WALT DISNEY

Transplanting Moss

Adding moss to your garden can be as simple as taking it from somewhere else and placing it in your desired area; this is called moss transplantation. Have some moss growing on a tree or rocks in your garden? Seen some growing in your neighbourhood? There's your start. Find patches of moss and carefully pull them up in large pieces (making sure you have permission to do so). The roots are short, so they're likely easy to remove. Then, bring the moss to the part of your garden where you want it to grow. Place the chunks down and press them firmly onto the surface. Pierce each piece with a stick to hold it down until it has time to root in. Spray regularly with water, and within weeks the moss will take to its new home and start growing and spreading on its own.

Growing Your Own

Moss, besides its many benefits, looks beautiful as a landscaping element; it's soft, dense, beautifully coloured, and it feels pleasant to touch. It also adds a natural look to a garden or piece of outdoor décor.

The following technique won't be practical for coating a large space or your entire garden, but if you want to strategically add some moss to your garden – perhaps on a statue, pot, tree, fountain, lawn décor, bench or around a specific garden bed – this could be a great solution.

You will need:

Plain yogurt
A small amount of native local moss
A blender
A wide paintbrush

Method

1. In your blender, combine the yogurt and moss. Use about 1 cup of yogurt for every tablespoon of moss.

2. Blend until the mixture is nice and smooth, then pour it into a bowl and head outside.

3. Use your paintbrush to spread the yogurt mixture onto the surface of your choice in a thick layer.

4. Put the item in a shady area (or choose a shaded area to spread the mixture), then mist daily with water and watch moss start to grow in a couple of weeks.

Conclusion

By now, I hope you've felt inspired to create your own wildlife haven – or perhaps just tried out a few projects in this book to elevate your outdoor space. At the very least, I hope you've realized that letting nature into your space to thrive and expand is easier than you might have first thought.

We're all trained to see gardening as a very precise, controlled endeavour; our lawns must be neatly trimmed, and our bushes tidily pruned. But that's only the expectation because we've made it so. By choosing to foster and welcome animals and natural growth in our outdoor spaces, we give the earth a little space to heal – and we also bring ourselves a huge amount of joy! Not only does creating something with your hands feel so good, but witnessing tiny moments in nature

provides us with an endless source of happiness.

So, play host to a nest of birds. Be a resting spot for sleepy hedgehogs. Give frogs and dragonflies places to lay their eggs and hunt for prey. Welcome minibeasts and remember that nature is one big cycle. Plant flowers just because butterflies love them. Support bees as they pass by on their journey to help create the food you eat. Whatever your intentions are – whether you want to give back to the planet, help some animals or just enjoy observing life outside your window – you're doing something wonderful.

The birds and bees (and their friends) will all thank you.

Notes

Image Credits

Front cover – background © W. Phokin/Shutterstock.com; top left © KPG_Payless/Shutterstock. com; top middle © Igor Podgorny/Shutterstock.com; top right © Juergen Bauer Pictures/ Shutterstock.com; bottom left © Lois GoBe/Shutterstock.com; bottom middle © Edita Medeina/Shutterstock.com; bottom right © Rudmer Zwerver/Shutterstock.com

Spine – ladybird © SuperArtWorks/Shutterstock.com

Back cover – background © W. Phokin/Shutterstock.com; background flowers © ZiaMary/Shutterstock.com; left © Coatesy/Shutterstock.com; middle © tomas.klacek/Shutterstock.com; right © hermaion/Shutterstock.com

p.3 – background © W. Phokin/Shutterstock.com; p.3 – top left photo © KPG_Payless/Shutterstock. com; p.3 – top middle photo © Igor Podgorny/Shutterstock.com; p.3 – top right photo © Juergen Bauer Pictures/Shutterstock.com; p.3 – bottom left photo © Lois GoBe/Shutterstock. com; p.3 – bottom middle photo © Edita Medeina/Shutterstock.com; p.3 – bottom right photo © Rudmer Zwerver/Shutterstock.com; pp.4–5 © Titus Group/Shutterstock.com; p.7 © Tongsai/Shutterstock.com; p.8 – left photo © Kuttelvaserova Stuchelova/Shutterstock.com; p.8 – middle photo © encierro/Shutterstock.com; p.8 – right photo © IanRedding/Shutterstock.com; pp.8, 24, 40, 54, 56, 72, 88, 102, 118 – background © ami mataraj/Shutterstock.com; pp.9, 22, 25, 32, 37, 41, 45, 51, 57, 59, 73, 80, 81, 89, 98, 103, 105, 115, 116, 117, 119, 128, 132 – background © Picsfive/Shutterstock.com; pp.10, 26, 42, 58, 75, 90, 120 © VerisStudio/Shutterstock.com; p.11 © burbura/Shutterstock.com; p.12 – left photo by Annie Spratt on Unsplash; p.12 – right photo © sasimoto/Shutterstock.com; p.13 photo by Rotem Vazan on Unsplash; p.14 © Tycson1/Shutterstock.com; pp.15, 19, 27, 34, 44, 46, 62, 64, 67, 79, 91, 95, 96, 107, 108, 109, 110, 113, 114, 123, 125, 133 – background © Paladin12/Shutterstock.com; p.15 – birdhouse illustration © NataliiaMiethe/Shutterstock. com; p.15 – birdhouse photo © Artem Kniaz/Shutterstock.com; p.16 © Erda Estremera/Shutterstock. com; p.15 © Nana_Hana/Shutterstock.com; p.17 © blew_s/Shutterstock.com; p.18 © JGade/Shutterstock. com; p.20 © Denys90/Shutterstock.com; p.21 © Claudine Silaho Weber/Shutterstock.com; p.23 – apricots and raspberries © Daria Ustiugova/Shutterstock.com; p.23 – plums © MarinaSM/Shutterstock.com; pp.23, 31, 33, 39, 51, 69, 86, 99, 101 – background © luceluce/Shutterstock.com; p.24 – left photo © Gardens by Design/Shutterstock.com; p.24 – middle photo © tony mills/Shutterstock.com; p.24 – right photo © rudiPro/Shutterstock.com; p.27 © Dan Freeman/Shutterstock.com; p.28 – left photo © Sophie McAulay/Shutterstock.com; p.28 – right photo © spoli/Shutterstock.com; p.29 © Laura Jarriel/ Shutterstock.com; p.30 © Thongchai /Shutterstock.com; p.31 © brajianni/Shutterstock.com; p.33 © Juergen Bauer Pictures/Shutterstock.com; p.35 © Gabriela Bertolini/Shutterstock.com; p.36 © Ka Yansh/ Shutterstock.com; p.38 © Nikilev/Shutterstock.com; p.39 © Nadya Krupina/Shutterstock.com; p.40 – left photo © Coatesy/Shutterstock.com; p.40 – middle photo by Andrew Coop on Unsplash; p.40 – right photo

If you're interested in finding out more about our
books, find us on Facebook at **Summersdale Publishers**
and follow us on Twitter at **@Summersdale.**

www.summersdale.com